A Brief History of Quantum Physics

Luís Orlando Emerich dos Santos

January 18, 2024

Contents

I	**History**	**9**
1	Max Planck and blackbody radiation	11
2	Einstein and the photoelectric effect	19
3	Rutherford and the nucleus of the atom	25
4	Bohr and the atom structure	31
5	Louis de Broglie and the wave mechanics	37
6	Werner Heisenberg and matrix mechanics	41
7	Erwin Schrödinger and the wave equation	47
8	Paul Dirac and the relativistic quantum mechanics	53
II	**Interpretations**	**61**
9	The Double Slit Experiment	63
10	The Copenhagen Interpretation	67

11 De Broglie-Bohm's Theory	**75**
12 The Many Worlds Interpretation	**79**
13 Final Remarks	**83**

Preface

In the late nineteenth century, many renowned physicists believed that there was nothing fundamentally new to be discovered, nothing to be changed in the physical conception of the universe. Of course, there were still some specific problems to be understood and gaps to be filled, but this could be done by applying the theories already developed: classical mechanics, thermodynamics, and electromagnetic theory. The turn of the century and the years that followed would show how far this was from the truth. The Theory of Relativity and Quantum Mechanics would profoundly alter the foundations of physics.

In this small book, I present the main moments and most important characters of the scientific revolution that was the development of Quantum Mechanics. Some important points were left out, which is inevitable in view of the intended concision. I believe, however, that with what remains it is possible to understand the plot and recognize the main characters. I have divided the book into two parts: in Part I, we have the development of the theory, from its beginning with Max Planck to the development of relativistic quantum mechanics with Paul Dirac; in Part II, I present the main interpretations of the theory.

This is a work written for the lay reader, who is used to hearing the word quantum and has no precise notion of its meaning. I avoided the use of equations and more technical subjects. The discussions presented in Part II, however, may be of interest to those who already have knowledge in the field. I'd like to express my gratitude to Prof. Juan Pablo de Lima Salazar for reading the manuscript and for the suggestions given.

Figure 1: Timeline of quantum physics, from Planck to Dirac.

Part I

History

Chapter 1

Max Planck and blackbody radiation

Figure 1.1: Max Planck

The history of quantum mechanics begins with Max Planck, born in 1858 in Kiel, northern Germany[1], the son of a law

---

[1]Germany as we know it did not yet exist. What there were small states that together formed the German Confederation. The unification of these states into a single nation occurred only in 1871.

professor. The family had an academic tradition, his grandfather and great-grandfather had been professors of theology. After studying in Munich and Berlin, he obtained the qualification to be a professor in 1880, presenting a thesis on thermodynamics. He began his academic career at Munich University. In 1885 he returned to his hometown and was appointed associate professor of Theoretical Physics at the University of Kiel. Finally, in 1889, he took up a position at the University of Berlin, where he would remain until the end of his career.

The years that interest us are the late 19th century, especially 1899. Planck was a full professor, was married, had two sons and two daughters. He was a respected, religious, and conservative citizen, and was about to start a revolution in the fundamentals of physics. To understand what led him to this, we must first have some knowledge of thermodynamics, waves, and electromagnetism. So, let's address these issues very briefly, then we will return to the late 19th century.

Have you ever wondered what heat is? What do we mean when we say a body is warmer or colder? This question was asked by various scientists in the seventeenth and eighteenth centuries. The main (wrong) theory was that heat would be a fluid, the caloric fluid. A body with more caloric fluid would be warmer and a body with less caloric would be colder. If a warm body was put in contact with a colder body, the caloric would flow from the hotter to the colder. It seemed like a good theory, but it had a flaw. How do you explain the heat that comes from friction? If we friction one body with another, both will warm up. Would the caloric fluid be coming out of nowhere? It wasn't a good

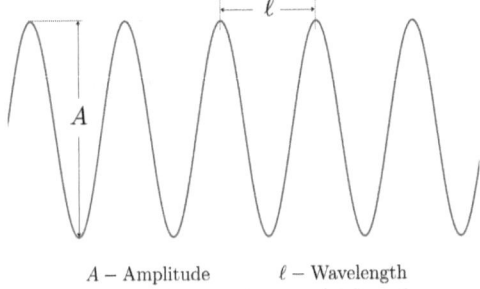

Figure 1.2: Simple (sinusoidal) wave.

explanation. Another theory (this one, correct) was that heat would be a form of energy, the same energy that is associated with movement. The movement of the particles that compose a body would manifest in the form of heat[2]. This movement could be the displacement, rotation, and vibration of atoms and molecules. By attrition of one body with another, we are transferring energy, we are moving the particles of solids, so they heat up.

Heat can be transferred by the exchanging of momentum[3] (movement) among particles. This can happen through collisions between the particles of a gas, or through the vibration of the molecules of a solid, which makes other molecules vibrate, and thus the heat is conducted. Heat can also be carried by the movement of a fluid in a process called convection. Another form of heat transfer is radiation, that is, through electromagnetic waves. That is the form that interests us.

---

[2] To be more precise we should say that heat is the transfer of energy, but such precision is not necessary for the following discussions.

[3] The momentum is defined as the product of mass by velocity.

Scottish James Clerk Maxwell was born in Edinburgh in 1831. At the age of fourteen, he published his first scientific paper, at eighteen he was already a recognized mathematician and continued studying natural phenomena, becoming the greatest physicist of the nineteenth century. In 1864, Maxwell unified all previous knowledge of electromagnetism into a set of equations, which today are known as Maxwell's Equations. By applying his equations, he predicted that electric and magnetic fields could propagate through space as electromagnetic waves that would travel at the speed of light. The light itself, according to Maxwell, would be an electromagnetic wave. He was right, light is an electromagnetic wave, this would be demonstrated experimentally a few years later.

When we talk about waves, we are talking about an oscillating movement that propagates in space. In the case of sound waves, the pressure oscillates and propagates, in the case of electromagnetic waves, the electrical and magnetic fields oscillate and propagate. A wave is characterized by its propagation speed, amplitude, frequency, and wavelength (see Figure 1.2). There is not much to say about the speed of propagation, it is the speed with which the wave propagates and depends only on the medium in which the wave propagates. Amplitude refers to the difference between the maximum and minimum values that occur in oscillations, in the case of a sound wave, is how much pressure varies, in the case of electromagnetic waves, is how much the intensity of the fields varies. The wavelength is the distance that the wave travels until it repeats itself. The frequency is the number of oscillations per unit time, or how fast these oscillations occur. These quantities are not independent, if a wave has a higher frequency it is because it has a shorter wavelength and vice versa. Waves are usually quite complex, formed by the sum of waves of various frequencies and amplitudes (Figure 1.3). But these complex waves can be broken down into simple waves, each with a single frequency and amplitude. When we represent the decomposition of a wave into its various frequencies we have the spectrum of the wave. In Figure 1.3 we have, in the upper part, a sound wave formed by the sum of waves of five different frequencies, in the lower part we have the spectrum of the wave, that is, how much of each frequency is necessary to form the wave of the upper part. The spectrum does not need to be discrete as in the figure, it can also be continuous. Electromagnetic waves can be produced – among other forms – by the accelerated movement of electrical charges, for example, when molecules vibrate. And here we talk about the heat again. When we heat a body the atoms and molecules that make up this body vibrate. That is, what we have are electrical charges in accelerated movement and, therefore, every body that has some heat – a temperature above absolute zero – emits electromagnetic waves. A body can also absorb electromagnetic waves, in this case, the phenomenon reverses itself and the body heats up. In everyday situations, bodies emit, absorb, and reflect radiation. All right, now we can go back to the end of the 19th century. One problem physicists were trying to understand was the relationship between a body's temperature and the spectrum of radiation it emits. Another way to understand the problem is to think about how the energy is distributed among the various frequencies. To simplify the problem they created an idealization, a black body, that is, an object that does not reflect any radiation

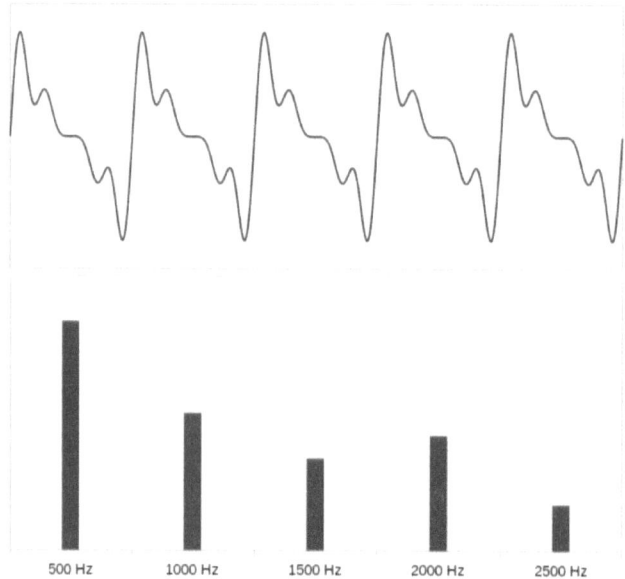

Figure 1.3: In the top we have a wave composed of the sum of simple waves. In the bottom we have the spectrum of the wave, that is, the frequencies that compose the wave.

only absorbs and emits. In this case, the spectrum of the emitted radiation will depend only on the body's temperature. But all theoretical attempts to obtain the spectrum had failed. The theoretical results – when they were not absurd – did not correspond to the experimental results. That's when Planck started to study the problem.

Planck imagined the black body to be composed of a set of oscillators[4]. It would be enough to apply the equa-

---

[4]We can think of oscillators as the molecules that make up the body.

tions of electromagnetism and deduce how the waves would be generated by the oscillators – which would be electrical charges oscillating. From there it would be possible to obtain the spectrum of the black body radiation for a given temperature. It seemed a safe path, but it did not lead to the solution. He then turned to the solutions proposed by other theoretical physicists. There was the Law of Vienn, which gave good results for the region of high frequencies, but failed at low frequencies and there was the Law of Rayleigh-Jeans which had the opposite behavior. From the two solutions, he was able to elaborate a mathematical expression compatible with the results of the experiments. It was a partial success because, although the expression obtained reproduced the experimental data, there was no physical basis to justify the expression. But now he knew where to go, he just needed to find his way. He returned to the oscillators model, testing hypotheses of how a set of oscillators, emitting and absorbing radiation, could lead to the mathematical expression he had deduced. It was, in his words, *the most strenuous work of my life.*

Finally, Planck managed to find a hypothesis that led to the expected result. The energy of the oscillators should be proportional only to the oscillation frequency[5] and this proportionality – this is the fundamental point – would be given through an integer. That is, for a given oscillation frequency, the energy could only be absorbed or emitted in *packages*, multiples of a fundamental quantity that he called quantum. It was a revolutionary and counterintuitive

---

[5]This was already contrary to the physics of the time because in classical physics the energy of an oscillator is proportional to the square of the amplitude of the oscillation.

hypothesis. Until then, in all known physical systems energy could assume any value, varying continuously. And now, to explain the spectrum of black body radiation, one had to assume that energy was quantized. The results obtained by Planck were published in December 1900 and were, for a time, ignored.

The physicists of the time – including Planck himself – did not consider the consequences of the quanta hypothesis, nor did they consider it to be fundamental. Although the solution found was in agreement with the experimental results, the quanta hypothesis would be only a temporary arrangement and, in the future, would be replaced by something more solid. That was what they think...

It took audacity to consider the quantization of energy as fundamental, and it took a genius to understand its consequences and broaden its scope. It took Albert Einstein.

Chapter 2

Einstein and the photoelectric effect

Figure 2.1: Albert Einstein

Albert Einstein was born in Ulm, southern Germany, in 1879. The family moved to Munich the following year and it was there that Einstein began his studies. At the age of sixteen, he moved to the city of Aarau, Switzerland, where he finished high school in 1896. He then began his mathematics and physics studies at the Zurich Polytechnic School. While Planck elaborated his ideas on black body radiation, Einstein finished his degree. Although he performed well academically, he did not achieve the desired teaching position. He worked as a substitute teacher in secondary schools and gave private lessons for two years, until, with the help of a friend, he got a job at the Patent Office in Bern, also in Switzerland.

It was a quiet job and in his spare time he could devote himself to the physical problems that interested him. He married a colleague from the Polytechnic School – Mileva Marić – and made friendships that would last a lifetime. To one of those friends he wrote in 1905:

> ... I promise you four papers [...] The first deals with radiation and the energy properties of light and is very revolutionary, [...] The second paper is a determination of the true size of atom... The third proves that bodies on the order of magnitude 1/1000 mm, suspended in liquids, must already produce an observable, random motion that is produced by thermal motion. Such movement of suspended bodies has actually been observed by physiologists, who call it Brownian motion. The fourth paper is only a rough draft at this point, and is an electrodynamics of moving bodies which employs a modification of the theory of space and time.[1]

Any of the four papers would be enough to give a physi-

---

[1] Einstein – His Life and Universe. Walter Isaacson – Simon and Schuster, 2007.

cist notoriety, but Einstein was not satisfied and, in addition to the promised papers, he also published, in the same year, a fifth paper in which he presented his ideas on the equivalence between mass and energy, the famous equation $E = mc^2$. Considering that our interest is the history of quantum mechanics, we will focus on the first of the promised papers, the revolutionary article on the energetic properties of light. Before, however, it is necessary to know a little about the photoelectric effect.

In the late 19th century, scientists who had been studying and experimenting with electromagnetism realized that metallic surfaces could eject electrons when illuminated. The phenomenon was observed by placing two metallic plates in a vacuum environment and subjecting the plates to a difference in electrical potential. When the plate with positive potential was illuminated the result was an electrical current between the plates – the photoelectric effect.

Varying the frequency and intensity of the light, two intriguing characteristics of the photoelectric effect were observed:

I - The energy of the ejected electrons was proportional to the frequency of the light incident on the metal. The higher the frequency of the light, the greater the energy of the electrons. On the other hand, for very low frequencies (for example, infrared light) electrons were not ejected from the metal, there was no electrical current, even when the intensity of the light was increased.

II - Keeping the light frequency constant (as long as it is not too low) and increasing the intensity, it was observed that the number of ejected electrons increased, that is, the electric current increased.

These results were difficult to explain by classical theory. Light, as we mentioned in chapter I, is an electromagnetic wave and the energy of an electromagnetic wave is independent of its frequency. However, what the experiments showed was the opposite, the energy that electrons acquired when ejected was proportional to the frequency and independent of the intensity of the light.

Einstein deciphered the puzzle by applying Planck's energy quanta hypothesis to phenomena involving electromagnetic waves. His hypothesis was that light would be composed of small particles – quanta of light, or photons, as they became known. If we imagine light as being composed of photons and the energy of photons as being proportional to the frequency of light, then what was mysterious about the photoelectric effect can be easily explained. Electrons are ejected from metal when they are hit by photons and the energy with which electrons are ejected depends on the energy of the photon that hits it. Thus, the greater the frequency of light, the greater the energy of the photons and therefore the greater the energy that electrons acquire. On the other hand, when the frequency is low, the energy of the photons is not sufficient to eject the electrons, no matter how many reach them. To increase the intensity of light would mean to increase the number of photons, not their energy, therefore it would only increase the number of electrons ejected – the measured electric current.

This explained the peculiarities of the photoelectric effect. But there was another problem. Several experiments showed that light behaved like a wave. How to reconcile these two different views? For Einstein, the wave behavior of light was an effect of the average behavior of a very large number of photons. In his 1905 paper, he wrote:

> The wave theory of light, which operates with continuous spatial functions, has proved to be excellent in describing exclusively optical phenomena and will probably never be replaced by another. However, we must keep in mind that optical observations refer to time averages and not instantaneous values ...

As we will see later, wave-particle duality is not exclusive to photons and is at the heart of quantum mechanics.

Although it explained the photoelectric effect simply, Einstein's paper was received with skepticism. Planck, who had introduced the quanta hypothesis, and had become an admirer of Einstein's work, would state in 1913 that with the quanta hypothesis of light, he had missed the target. But, despite the initial skepticism, the experimental evidence grew until, in 1921, Einstein was awarded the Nobel Prize in physics for his work on the photoelectric effect.

As the enigmas of blackbody radiation and the photoelectric effect were deciphered, many more emerged. The study of the structure of matter, that is, of the atom, seemed to be a source of enigmas. This will be the subject of our next chapter.

Chapter 3

Rutherford and the nucleus of the atom

Figure 3.1: Ernest Rutherford

Ernest Rutherford – who would come to be known as the father of nuclear physics – was born in 1871 in a rural community near the city of Nelson, New Zealand. He was the fourth of twelve children of James and Martha Rutherford, who had emigrated from Scotland. After finishing high school with an outstanding performance in all disciplines, he was awarded a scholarship to study at the University of New Zealand in the capital Wellington, where he graduated in physics and mathematics. In 1894, he obtained a scholarship to continue his education at the University of Cambridge, England. There he studied under the guidance of J. J. Thomson – who, as we will see below, is the discoverer of the electron.

In 1897, he moved to Montreal, Canada, accepting the offer of a teaching position at McGill University. It was at McGill University that Rutherford would do the work on the disintegration of radioactive substances[1] that would give him the 1908 Nobel Prize in Chemistry. That year, however, he was already working at Manchester University, having returned to England the year before. It was at Manchester University that Rutherford conducted experiments that enabled him to discover the nucleus. Before we look at these experiments, let's take a look at the development of the history of the atom.

The word atom has its origin in ancient Greece and means indivisible. The idea that all matter would be composed of indivisible elements – the atoms – was proposed by the philosopher Leucippus and systematized by his disciple Democritus. With the development of chemistry, the atom concept was used to explain why elements always react in a ratio of whole numbers. John Dalton – considered the pioneer of modern atomic theory – proposed in 1803 that chemical elements were composed of atoms of a single type, and chemical compounds, by combining atoms of dif-

---

[1]Radioactive substances are unstable chemical elements, which disintegrate into other elements and emit radiation.

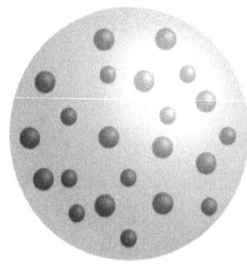

Figure 3.2: Atomic model proposed by J. J. Thomson. The small spheres would be the electrons.

ferent types. Although it was widely used by chemists, the atomic hypothesis was slow to be accepted by physicists, whose resistance was only finally overcome after Einstein's work on the Brownian movement (1905), which allowed the determination of the mass and size of atoms.

Although there was resistance, some physicists proposed theories and carried out experiments considering the matter as being composed of atoms. And what the experiments would show is that atoms are not indivisible, but composed of even smaller particles.

In the year 1897, J.J. Thomson performed experiments with cathode rays. Cathode rays appear when metal plates are placed in a glass ampoule filled with gas at low pressure and subjected to high electrical voltage[2]. At the time, there was controversy about the composition of cathode rays. Were they particles or waves? By subjecting the cathode rays to electric and magnetic fields, J.J. Thomson was able to demonstrate that they were composed of particles

---

[2]Antique televisions are cathode ray tubes.

with a negative electrical charge – he had discovered the electron. Since it was possible to generate cathode rays using plates of the most diverse metals it was possible to deduce that electrons were part of the constitution of all matter. On the other hand, as matter in normal situations has no electrical charge – the atoms are neutral – it was possible to deduce that the rest of the atom would be composed of matter with a positive charge. Hence, in 1904 Thomson proposed a model in which the electrons are embedded in a sphere of positive charge (Figure 3.2). This model would have a short life, it would be abandoned due to experimental results obtained by Ernest Rutherford.

During his nine-year stay in Canada, Ernest Rutherford – among other achievements – had discovered alpha, beta, and gamma radiation1. In 1909, Rutherford and his assistants Hans Geiger and Ernest Marsden built an experiment in which an alpha particle beam reached a very thin gold plate (see Figure 3.3). If the model proposed by Thomson was correct, the particles would be slightly deflected by the gold atoms.

Although it adequately explained the results of the gold leaf experiment, Rutherford's model had a serious flaw. As we saw in Chapter 1, electrical charges in accelerated motion emit electromagnetic waves. The electrons orbiting the nucleus should, therefore, emit electromagnetic waves continuously. But by emitting radiation, the electrons would be losing energy. As a consequence, the electrons would tend to decrease the radius of their orbits until they reach the nucleus, and the atom would cease to exist. This enigma would persist until a young Dane named Niels Bohr came to work with Rutherford in Manchester.

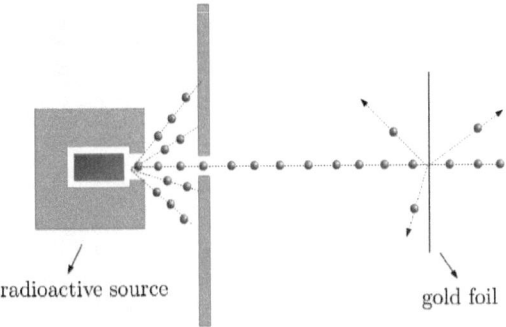

Figure 3.3: Rutherford's experiment. A beam of radiation hits and is spread by a gold foil.

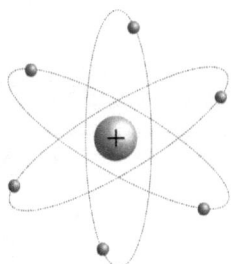

Figure 3.4: Atomic model proposed by Rutherford, with the positive nucleus and electrons orbiting as planets.

Chapter 4

Bohr and the atom structure

Figure 4.1: Niels Bohr

Niels Henrik David Bohr was born in Copenhagen, Denmark, in 1885. He was the second son of Christian Bohr,

Figure 4.2: Visible part of the spectrum of the hydrogen atom.

a professor of physiology, and Ellen Adler Bohr, who came from a prominent family of Jewish bankers. He studied physics at the University of Copenhagen and, while still a student, was awarded a gold medal from the Royal Danish Academy of Sciences and Letters for a paper on the surface tension of liquids. Following his academic career, he obtained a master's degree in 1909 and a doctor's degree in 1911. After a stay in England, he would publish, in 1913, three fundamental articles on the atomic structure. We will return to the three articles below, but, before, we must know a little about atomic spectra.

Spectroscopy is the study of the interaction – emission and absorption – between electromagnetic radiation and matter. As we mentioned in Chapter 1, when we heat a material its atoms and molecules vibrate, emitting electromagnetic radiation. We can determine the spectrum[1] of this radiation by passing it through prisms. If the heated substance is a simple element like hydrogen, the spectrum will not be continuous like a rainbow, but a set of spectral lines (Figure 4.2).

The atomic models proposed by J.J. Thomson and Rutherford could not explain the spectral lines. In the case of the

---

[1] Remembering, spectrum is the set of frequencies that make up the radiation.

first model, because the electrons were not in motion. In the case of the second, because the energy would be emitted in a continuous spectrum.

In 1911, after completing his doctorate, Bohr obtained a scholarship from the Carlsberg Foundation and left for England to study with J.J. Thomson at Cambridge. It was an unproductive period, Bohr had difficulties with the language and did not get along well with Thomson, who, according to him, was a difficult person to talk to and unable to accept criticism. Having met Rutherford on a trip to Manchester in November, Bohr applied and got a transfer to work with his team. His goal was to learn about radioactivity.

At first, Bohr tried to work with experiments involving the emission and absorption of alpha particles but soon realized that experimental work was not his vocation. Returning to the theoretical work, he began to think about the problems of Rutherford's atomic model. He intended to apply Einstein and Planck's quantization concepts to explain the stability of atoms. His investigations continued from 1912 until 1913, when, having returned to Copenhagen, he published his results.

As we mentioned, the electromagnetic theory predicted that electrons in circular orbits would emit radiation, thus losing energy and collapsing in the nucleus. The atom would be unstable. But the hydrogen atom is stable. Therefore, Bohr concluded, the electromagnetic theory – especially the emission of electromagnetic waves – could not be applied to the atom, just as it could not be applied to explain the photoelectric effect or the spectrum of the black body radiation. Bohr postulated that there would be a set of stable orbits – stationary states – in which the electrons would be confined. In these stationary states, electrons would not emit radiation. Bohr was saying that the orbits – and, therefore, the energy associated with these orbits – were quantized. Each orbit would be associated with an energy level. The lowest energy level was called the fundamental state, which is when the electron is closest to the nucleus. Based on Einstein's work, Bohr proposed that electrons could emit and absorb radiation in the form of photons. An electron could absorb a photon and jump into a more outer orbit. Similarly, an electron that was not in the fundamental state could jump into a more internal orbit, emitting a photon. The energy of the photon would be equal to the difference between the energy levels of the orbits. As the photon's energy is proportional to its frequency and the energy varies in a discontinuous manner, the same occurs with the frequency. In other words, instead of a continuous spectrum, we would have spectral lines.

With his postulates and making a parallel between the quantum and classical quantities[2], Bohr was able to predict what the energy levels of the hydrogen atom would be – how the spectral lines would be distributed. It was also possible to calculate what would be the energy necessary to remove an electron from the hydrogen atom[3]. Both results were corroborated by experiments. His theory of the structure of the atom was acclaimed and boosted the study of quantum physics. In 1922, Bohr was awarded the Nobel Prize in Physics *for his services in investigating the structure of the*

---

[2]Bohr elaborated the correspondence principle, according to which quantum quantities should tend to classical quantities.

[3]The minimum energy to take an electron out of an atom is called ionization energy.

atom and the radiation it emitted and became the mentor of a generation of theoretical physicists who would travel to Copenhagen in search of inspiration.

Bohr's theory would be enhanced by Arnold Sommerfeld – a renowned German physicist – and became known as Bohr-Sommerfeld's quantization rules. But despite the success achieved in explaining the structure of the hydrogen atom, all attempts to extend the theory to other atoms proved fruitless. A more comprehensive and fundamental theory was needed.

Chapter 5

Louis de Broglie and the wave mechanics

Figure 5.1: Louis de Broglie

Louis-Victor Pierre Raymond de Broglie was born in 1892 in Dieppe, France, the youngest son of a wealthy aristocratic family. His father, Victor, was the Duke of Broglie. His brother Maurice[1], eighteen years older, was already a

---

[1]Maurice de Broglie was a renowned physicist who worked with

scientist when Louis de Broglie became interested in science, after graduating in history at the age of eighteen. He started to frequent his brother's laboratory and studied the works of Henri Poicaré, Hendrik Lorentz, Paul Langevin, Ludwig Boltzmann, Josiah Gibbs, Albert Einstein and Max Planck, graduating in science in 1913. In the same year he joined the French army and started working with radiotelegraphy. He would remain in the army throughout World War I, taking the opportunity to study electronics and deepen his knowledge about electromagnetic waves.

With the end of the war, de Broglie resumed his studies in physics, conducting experiments with his brother in studies on X-rays and the photoelectric effect. Returning to theoretical physics, he published, between the years 1922 and 1923, five articles in which he developed the principles of wave mechanics. These articles would be the basis of his Ph.D. thesis, defended in 1924.

We commented in chapter II about the wave-particle duality, the strange fact that light presents both an undulatory and corpuscular behavior. Louis de Broglie postulated that not only photons but all forms of matter should have corpuscular and wavy properties, thus associating waves to particles. In his words: *...in my conversations with my brother we always came to the conclusion that in the case of X-rays there are always particles and waves, so suddenly... I had the idea that this duality should be extended to material particles, especially to electrons.*[2]

His starting point was Einstein's work on the photoelec-

X-rays and was even nominated for the Nobel Prize.

[2]Louis de Broglie's interview with T.S. Kuhn, A. George and T. Kahan.

tric effect and the equivalence between mass and energy. Einstein had related the photon energy to the frequency of light and had demonstrated the equivalence between mass and energy through the famous equation $E = mc^2$. Joining the two ideas it is possible to associate a frequency to a resting particle: its mass implies certain energy, which in turn implies a certain frequency. For the description to be complete it was necessary to determine a wavelength. Once again he resorted to Einstein's works, using the Theory of Special Relativity, he managed to relate the mass of the particle to a wavelength[3].

After associating a wave to the electron, Louis de Broglie could deduce the orbits predicted by Bohr by imposing the condition that the electron's trajectory around the nucleus should contain an integer number of wavelengths, in which case the waves are stationary (see Figure 5.2). This gave an explanation of why electrons can only assume certain orbits - a fact that, before, seemed something totally arbitrary. In the conclusion of his doctoral thesis, de Broglie would write: *...we believe that this is the first physically plausible explanation proposed for the Bohr-Sommerfeld stability conditions*[4].

A direct consequence of the theory proposed by de Broglie is the existence of waves of matter, something that was experimentally proven in 1927 by C. J. Davisson and L. H.

---

[3]For those who want to check this deduction, we suggest reading Modern Physics by Raymond Serway, Clement J. Moses and Curt A. Moyer.

[4]Louis de Broglie. Recherches sur la théorie des Quanta. Physique [physics]. Migration - université en cours d'affectation, 1924. Français. tel-00006807

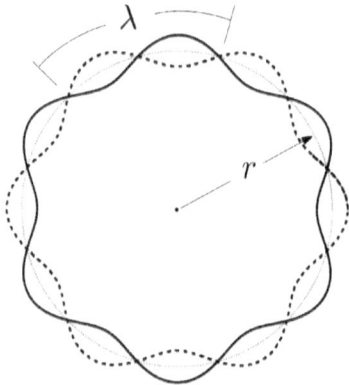

Figure 5.2: The orbit radius of the Bohr stationary states corresponds to an integer of wavelengths (λ). In the figure there are six wavelengths.

Germer in the United States and by George P. Thomson in Scotland. Soon after, in 1929, Louis de Broglie was awarded the Nobel Prize in Physics *for his discovery of the undulatory nature of electrons.*

Some points, however, were still unclear. There was no adequate interpretation of what the waves of matter were, and an equation describing the behavior of these waves was missing. The deduction of the wave equation would be left to a brilliant Austrian physicist who oscillated between physics and philosophy, Erwin Schrödinger. But first came the contribution of a young German physicist. That is what we will see in the next chapter.

Chapter 6

Werner Heisenberg and matrix mechanics

Figure 6.1: Werner Heisenberg

Werner Karl Heisenberg was born in Würzburg, southern Germany, in 1901. He was the youngest of the two sons of August Heisenberg and Annie Wecklein. August was at the time a teacher of Greek language and literature in secondary schools. In 1910, the family moved to Munich, where August took up the position of professor of Greek philology at the University of Munich. The following year Werner joined Maximilians-Gymnasium – a secondary school that was run by his maternal grandfather – impressing his teachers by his early mastery of mathematics. He also studies music, even considering a career as a pianist. The class routine is disrupted by the outbreak of World War I in 1914. With more free time, he studies, on his own, differential and integral calculus. With the end of the war, in 1918, Germany enters a time of political turbulence. In Munich, the communists seize power and try to impose a Soviet-style republic. Werner enlists government troops that regain control of the city. When the situation returns to normal, he joins the German Youth Movement, organizing skiing and mountaineering excursions.

After finishing secondary school, Heisenberg enters the University of Munich, where he studies under the guidance of Arnold Sommerfeld - one of the exponents of quantum mechanics in development. At that time he made friends with Wolfgang Pauli – a prodigy who at eighteen had published an article on Einstein's Theory of General Relativity and who would be known for the formulation of the Exclusion Principle. This friendship was maintained through the years despite personality differences: Pauli was an avid nightlife frequenter who slept late and missed classes, Heisenberg was the opposite, a dedicated student who woke up early and enjoyed outdoor walks.

In 1923, Heisenberg defended his doctorate with a thesis on hydrodynamics. The defense didn't come out as expected, Heisenberg was harshly criticized by one of the members of the board – Wilhelm Wien[1] – for his limited

---

[1]Wilhelm Wien has become known for elaborating the Wien Law

knowledge of experimental physics. Despite the criticism, he got the approval and, upset by the episode, moved on to the University of Göttingen to work with Max Born. In the following years, he would divide his time between Göttingen and Copenhagen, where he would collaborate with Niels Bohr.

The conversations with Bohr, Born, and Pauli, and the attempts to understand the structure of the atom end up convincing him that an adequate theory would only be possible if it were restricted to what is effectively observed – the so-called observables – abandoning classical concepts such as the orbits of electrons. In May 1925, with a strong hay fever crisis, he decided to isolate himself in the small island of Helgoland, in the North Sea, in search of a faster recovery. Alone, without distractions, he concentrates on the problem of describing the atom using only observables.

The atom is only observed through the spectral lines, characterized by frequency and intensity. Spectral lines appear when atoms emit photons, that is, when electrons change their energy level – or orbit, in the Bohr model. Each change of energy level is characterized by two numbers, corresponding to the initial and final energies. Studying the simplest problem of a single electron, following the same reasoning, he organized the possible energies in rows and columns, associating to each term an amplitude and a frequency. Starting from the amplitudes he was able to calculate the intensity of the spectral lines[2]. Although he did

---

that correctly predicts the spectrum of black body radiation for high frequencies but fails at low frequencies.

[2]For those who want to see in detail how Heisenberg developed his reasoning I suggest reading "Heisenberg, Models, and the Rise of

not know, the mathematical operations he found necessary correspond to the operations of matrix algebra[3]. Enthusiastic about the results, Heisenberg wrote a manuscript and sent it to Born for publication.

One point that intrigued Heisenberg was that the rules he used to multiply his number arrangements implied a multiplication operation that was non-commutative, that is, the order of factors changed the result. Born and his assistant Pascual Jordan soon recognized that the operation was the multiplication of matrices. Non-commutativity was not really a problem, if the matrices represented measurement processes it was natural that the order was relevant. The three – Heisenberg, Born, and Jordan – then published two papers that would be the foundation of what became known as matrix mechanics – one of the versions of quantum mechanics. Wolfgang Pauli would be the first to apply the new mechanics to deduce the spectrum of the hydrogen atom, finding the correct solution.

It seemed that the riddles of the atomic scale were being deciphered. But the cost was high! It was necessary to give up trying to visualize the structure of the atom and apply mathematical concepts that were foreign to the everyday life of physicists.

It was then that – a little after the publication of the papers that elaborated the matrix mechanics – Erwin Schrödinger published his first papers providing the basis for the other version of quantum mechanics, the wavy mechanics, as will

---

Matrix Mechanics" by Edward MacKinnon - Historical Studies in the Physical Sciences - Vol. 8 (1977), pp. 137-188.

[3]The term matrix is used in mathematics to designate an arrangement of numbers, symbols, or expressions in rows and columns.

be seen in the next chapter. Before, however, it is necessary to mention another great achievement due to Heisenberg, the Uncertainty Principle.

Heisenberg was in Copenhagen in 1927 when he developed and published what he called uncertainty relations. He argued that physical quantities can only be defined when they can be measured. So there is no point in talking about the position or velocity of a particle when there is no way – even, theoretical – to measure these quantities. As an example, he considered the measurement of the position of an electron. If the measurement is made through something like a microscope, it would be necessary for a photon to reach the electron for the position to be determined, and the accuracy of the measurement would depend on the energy of the photon, the greater the energy the greater the accuracy. However, a photon hitting an electron will change its speed, and the greater the photon's energy, the greater the change will be. The consequence is that the *position and velocity magnitudes could not be determined simultaneously*. Then, in the same paper, he extended the relationships of uncertainty to other pairs of measures, such as energy and time. We will also comment on the Uncertainty Principle in the context of interpretations of quantum mechanics.

Chapter 7

Erwin Schrödinger and the wave equation

Figure 7.1: Erwin Schrödinger

Erwin Rudolf Josef Alexander Schrödinger was born in Vienna, Austria[1], in 1887. He was the only son of Rudolf Schrödinger and Georgine Emilia Brenda Schrödinger. Rudolf

---
[1]At the time Austria was part of the Austro-Hungarian Empire.

ran a small company, was an educated man, who studied chemistry and botany, and published papers on the phylogeny of plants. Georgine was of English descent and Erwin grew up speaking English and German. He studied at home until he was eleven when he joined the Gymnasium - the secondary school. His intellectual interests ranged from scientific subjects to German poetry. In 1906 he entered the University of Vienna, where he demonstrated the same level of excellence he had demonstrated in secondary school. During this period, he was strongly influenced by Fritz Hasenöhrl, who had succeeded Ludwig Boltzmann[2] in the chair of theoretical physics at the University of Vienna. In 1911, Schrödinger became Franz Exner's assistant, helping with experimental physics classes, and three years later, in early 1914, he obtained the qualification to lecture at the university.

A few months after Schrödinger obtained his qualification, World War I began. Summoned, he serves as an artillery officer. After participating in missions in the Tyrol and Budapest, he returns to Vienna with the assignment to teach an introductory meteorology course for anti-aircraft defense officers. During this period, despite the war, he managed to study and publish two papers on the Theory of General Relativity, which Einstein had finished in 1915.

With the end of the war, Schrödinger returns to the University of Vienna, but, remains for a short time. In 1920, he accepts an invitation to be an assistant professor at the University of Jena in Germany. After marrying Annemarie Bertel, he moved to Jena and remained there for less than

---

[2]Ludwig Boltzmann, one of the greatest theoretical physicists of the 19th century, committed suicide in 1906.

a year. He then holds the position of associate professor in Stuttgart, Germany, and a full professor in Breslau, Poland. Finally, in September 1921, he became a professor at Zurich University, where he would remain for six years.

Soon after settling in Zurich in the late months of 1921, he was diagnosed with suspicion of tuberculosis. For his recovery he chose the town of Arosa, a ski resort near Davos in the Swiss Alps – the couple would remain there most of 1922. In Arosa, he continues his research and publishes two articles. In one of them, he anticipates results that would be published by Louis de Broglie in 1924.

Schrödinger's personal and professional lives were complicated in 1925, he and Annemarie had extramarital affairs, and at the age of 38, all that he had published would yield him no more than a footnote in the history of physics. In October, he learned of Louis de Broglie's thesis through a paper by Einstein. He got a copy of the thesis and, at the request of his colleague Peter Debye[3], prepared a lecture, which was presented in November. At the end of the presentation, Debye would have commented that to properly speak of a wave it was necessary to have an equation that described the wave. This sounded like a challenge for Schrödinger. Just before Christmas, he left for a new season in Arosa. He took his notes on Broglie's thesis with him and was accompanied not by his wife, Annemarie, but by a lover whose identity is unknown to this day. When he returned in January, Schrödinger had deduced the wave equation.

The notebooks left by Schrödinger indicate the path he

---

[3]Peter Debye was an important North American physical chemist. He was awarded the 1936 Nobel Prize in Chemistry for his contributions to the knowledge of molecular structures

followed when deducing the equation. He took as his starting point the wave equation, known by physicists for describing sound waves and electromagnetic waves. In his first attempt, he introduced in the wave equation the relations between mass and frequency obtained by Broglie from the Theory of Special Relativity. The result was a relativistic wave equation[4]. Schrödinger then applied this wave equation to the hydrogen atom, but the results obtained were incorrect. This first attempt would only be published in the second half of 1926. In a second attempt, Schrödinger used as a starting point the Hamiton-Jacobi theory – which is a more abstract and mathematically advanced form of Newton's Laws – and did not use any relativistic relationship. The application of this new equation to the hydrogen atom led to the correct energy levels.

In the first months of 1926, Schrödinger published six papers on which he founded the wave version of quantum mechanics. His results were greeted with enthusiasm by physicists. Now it was possible to solve the problems on the atomic scale without the need to resort to the transcendental matrix algebra.

But there was a problem. If before there were no theories that could be applied comprehensively on the atomic scale, now there were two, and both theories seemed to be correct, although the mathematical form was very distinct: on one side there was the non-commutative matrix algebra, on the other, partial differential equations. Moreover, the theories led to completely different interpretations. For Schrödinger, the electron spreads like a wave of matter in orbit around

---

[4]Schrödinger deduced one of the forms from Klein-Gordon's equation, which would be published several months later.

the nucleus. Heisenberg, on the other hand, argued that it was not possible even to talk about orbits, but only about changes in the energy levels of electrons.

Wave mechanics was not very well received by Heisenberg, in a letter to Pauli he wrote: *The more I think of the physical portion of Schrödinger's theory, the more repulsive I find it.* Note that his rejection is mainly to what he calls the physical portion of the theory. Heisenberg rejected the possibility of visualizing the atom or referring to unobservable quantities. His position is philosophical, not scientific, and will be incorporated into the orthodox interpretation of quantum mechanics, as we shall see.

Despite Heisenberg's objections, wave mechanics was well received and embraced by most physicists – and for the same reasons that led Heisenberg to find it repulsive. Still, in June 1926, Schrödinger mathematically demonstrated the equivalence between the two theories. But, one question remained: What is the physical meaning of the wave function? Although it is possible to think of the electron as a wave of matter in the case of the atom, this interpretation fails in the case of a collision between an electron and an atom. In this case, the electron behaves like a particle. Born – one of the founders of matrix mechanics – applied Schrödinger's Equation to analyze this problem. The interaction results in a superposition of waves. It was not possible to say which direction the electron would go after the collision. Based on the classical theory, in which the square of the wave amplitude is associated with wave intensity, he proposed that the square of the wave function corresponded to a probability. Thus, although he could not state what direction the electron would take after the collision, he could assign a probability to each direction. There is much more to say about this interpretation, but we will postpone this discussion for the second part of the book.

While matrix mechanics was being developed in Göttingen, a young doctoral student at Cambridge University, inspired by Heisenberg's first paper on matrix mechanics, had developed his own version of the theory. That's what we'll see in the next chapter.

Chapter 8

Paul Dirac and the relativistic quantum mechanics

Figure 8.1: Paul Dirac

Paul Adrien Maurice Dirac was born in 1902 in Bristol, England. He was the second son of Charles Adrien Ladislas

Dirac and Florence Holten, who would still have a daughter. His father, Charles, was a Swiss immigrant who taught French and registered his children as Swiss citizens - only at seventeen would Paul become a British citizen. His mother had been a librarian before the marriage. Remembering his childhood years, decades later, Dirac would claim he never had a childhood. His father was authoritarian and distant, he had no friends and the family lived in self-imposed social isolation. Such a childhood would influence the adult, who would become an introverted and taciturn person.

Dirac did his elementary schooling at a small school near his family's home and then joined the Merchant Venturers' Technical College, where his father was a teacher. Finishing secondary school with excellent performance, Dirac was awarded a scholarship to Bristol University, starting his studies in electrical engineering in 1918. He spends most of his days isolated in the library and studies, on his own, Einstein's Theory of Special and General Relativity. Three years later he would graduate with honors and get a scholarship to continue his studies at Cambridge University. The amount of the scholarship, however, was not enough to maintain him at Cambridge. Without being able to get a job as an engineer, he eventually returned to the University of Bristol for a second degree, now in mathematics. During the course, he deepened his study of classical mechanics and took extracurricular courses on atomic theory. In two years he finishes the course and is awarded a government scholarship which, added to the scholarship he had obtained from the University of Cambridge, allows him to start his doctorate.

In Cambridge, Dirac would be known for his quietness

and his literal mind. A colleague, years later, would recall that when asked about some scientific question: *He looks five minutes at the ceiling, five minutes at the window, and then says 'Yes' or says 'No'. And he's always right.* In the cafeteria where the students gathered for dinner, Dirac was invariably silent. It is said that a colleague, to pull conversation, would have commented, *It's a little rainy today, isn't it?* Dirac got up, went to the window, and after returning and sitting down said, *Not at the moment.* In any case, Dirac's talent did not go unnoticed by his advisor, Ralph Fowler, who was one of the few English physicists up to date with the advances in quantum mechanics.

In September 1925, Fowler sent Dirac the paper Heisenberg had written in Helgoland, with the principles of matrix mechanics. On the front page, Fowler wrote, *What do you think of this? I'll be happy to hear it.* Dirac had a limited knowledge of German, which made it difficult to understand the text. In addition, he considered Heisenberg's approach – restricted to observable – very complicated and artificial. He left the manuscript aside for a week, then resumed his reading. The non-commutativity of the elements of the theory, which had intrigued Heisenberg, caught Dirac's attention. He had a habit of taking long walks in rural Cambridge and it was on a Sunday on one of these walks that he made his first major discovery.

Non-commutativity means only that if we multiply an element A by an element B we will get a different result from B multiplied by A. Dirac then noted that the difference AB - BA also appeared in classical mechanics through an operation known as *Poisson brackets*[1]. With this, it was pos-

---

[1] The equations of classical mechanics - in Hamilton's formalism -

sible to have a correspondence between the classical quantities and the quantum mechanics quantities. He'd been studying classical mechanics for some time and couldn't remember the details of Poisson's theory. Then he hurried back from his walk, but since it was Sunday and the libraries were closed, he had to wait until Monday to consult books on classical mechanics and verify the correction of his idea. On Monday morning he consulted the books in the library and saw that his conjecture made perfect sense. This was followed by weeks of hard work, in which Dirac developed the entire mathematical basis of his theory in analogy to classical theory. In November, Dirac handed Fowler his manuscript entitled *The Fundamental Equations of Quantum Mechanics*, which was sent to the Royal Society for publication. He also sent a handwritten copy to Heisenberg.

The proposed theory was more comprehensive, more ambitious, and more abstract than the theory Heisenberg had formulated in his paper which had inspired Dirac. From the classical analogy with Poisson's parentheses, Dirac formulated an equation of motion, that is, an equation that allowed to calculate the temporal evolution of a system.

Some of the results that Dirac had obtained had already been deduced and published in the paper written jointly by Heisenberg, Born, and Jordan, but this did not diminish the importance of what he had accomplished. In the letter Heisenberg sent to Dirac thanking the manuscript, he wrote:

> ... I hope you are not disturbed by the fact that indeed parts of your results have already been found here some time ago and are published in-

---

can be deduced from Poisson brackets.

dependently here in two papers- one by Born and Jordan, the other by Born, Jordan, and me – in Zeitschri.ft for Physik. However. because of this your results by no means have become less important [unrichtiger]; on the one hand, your results, especially concerning the general definition of the differential quotient and the connection of the quantum conditions with the Poisson brackets, go considerably further than the just mentioned work; on the other hand, your paper is also written really better and more concisely than our formulations given here.

The paper did not receive immediate recognition – it was too abstract for that – but, gradually, Dirac's name came to be known among physicists in the area.

After defending his doctorate, he published another very important paper in August 1926. When studying wave mechanics he realized that Schrödinger's equation, when applied to particles, allowed two types of solutions – symmetrical and anti-symmetrical – that led to distinct behaviors. The particles whose solutions were symmetrical obeyed a statistic developed by Satyendra Nath Bose and improved by Einstein, and which became known as Bose-Einstein Statistics. The anti-symmetric solutions were first studied by Enrico Fermi, in addition to Dirac himself, thus giving rise to Ferm-Dirac Statistics. The particles were then divided into two classes: the bosons and the fermions[2]. This work was immediately hailed as a major contribution to

---

[2]Wolfgang Pauli later showed that the bosons have integer spin and the fermions, semi-integer.

quantum mechanics, although it was considered extremely difficult. After reading the paper, Schrö dinger wrote in a letter to Bohr: *Dirac has a completely original and unique method of thinking, which – precisely for this reason – will yield the most valuable results, hidden to the rest of us. But he has no idea how difficult his papers are for the normal human being.*

Dirac's main contribution was yet to come. In 1926 there were still some important puzzles to be deciphered. The observation of the spectral lines of the hydrogen atom in high resolution showed that they were divided into thinner lines. The fine structure of the spectral lines was supposed to be the consequence of relativistic effects, which were not captured by Schrödinger's Equation. Another enigma was the electron spin – spectral lines were altered when atoms were subjected to magnetic fields, suggesting that electrons must have an intrinsic magnetic field, which was called spin[3]. The incorporation of spin in quantum mechanics – in the Schrödinger Equation or in matrix mechanics – was something that seemed quite artificial. Finally, there was the fact that Schrödinger's Equation was incompatible with the Theory of Special Relativity. Everything indicated that these problems were related.

From his student days, Dirac was fascinated by the Theory of Relativity. It was therefore quite natural for him to deduce a relativistic version of Schrödinger's Equation. When he commented on his intention with Bohr, he was discouraged by the comment that this problem had already been solved. Indeed, Oskar Klein and Walter Gordon had

---

[3]Spin is the term used to denote the intrinsic magnetic moment of particles.

published a relativistic version of Schrödinger's Equation, which became known as the Klein-Gordon Equation. As we saw in the previous chapter, the first to deduce such an equation was Schrödinger, although he was not the first to publish it. Dirac himself had also deduced the same equation independently but did not attach much value to it. The problem with the Klein-Gordon Equation is that, although it can be applied in some cases, it fails when applied to the electron[4].

In the Theory of Relativity space and time are treated in the same way, that is why we speak of a space-time continuum. This does not occur in Schrödinger's Equation, in which the variations in time occur distinctly from the variations in space[5]. In previous attempts to obtain a relativistic equation, Klein, Gordon, Schrödinger, and Dirac himself changed the way temporal variations appear in the equation. But Dirac was not happy with this approach. He decided to change the way variations in space appear in the equation. His interest was aesthetic, he was not concerned with the problem of the electron spin or the fine structure of the spectral lines of hydrogen, his concern was the form of the equation. It was mainly a mathematical work, something that suited his inclinations. As he would say, years later, *I think it is a peculiarity of mine that I like to play with equations, just looking for beautiful relationships that may not have any physical meaning. Sometimes they do.* In this case, when he found the relationships he wanted, they

---

[4]Klein Gordon's equation correctly describes only spin-zero particles.

[5]Schrödinger's equation relates a first derivative in time with a second derivative in space.

had a lot of meaning.

By applying the deduced equation for the case of an electron in an electromagnetic field the spin would appear correctly, without the need for any additional change. Also, the fine structure of the spectral lines of hydrogen appeared as a solution of the equation, which became known as the Dirac Equation. The paper with his main contribution to quantum mechanics was published in February 1928, with the title *The Quantum Theory of the Electron*.

Dirac's equation had a peculiarity that did not go unnoticed. When applied to the electron it allowed not one, but two solutions. One solution corresponded to the positive energy levels of the electron observed in the experiments, but there was also a solution that generated negative energy levels. Was there any physical meaning in the solution? Dirac examined the problem for three years, proposing solutions that proved to be faulty, until he proposed in 1930 that negative energy levels could only be explained by the existence of an anti-electron: a particle with the same mass as the electron, but with a positive electric charge. Two years later, the anti-electrons – which became known as positrons – were detected in cosmic rays.

Paul Dirac shared with Erwin Schrödinger the Nobel Prize of 1933 *for the discovery of new and productive forms of atomic theory.*

Part II

Interpretations

Chapter 9

The Double Slit Experiment

Thomas Young has already been described as the last man who knew everything. He spoke fourteen languages, practiced medicine, helped to decipher Egyptian hieroglyphics, and conducted experiments that were essential to the development of physics. In 1803, he presented to the Royal Society his most famous experiment, the double-slit experiment.

The purpose of the experiment was to demonstrate the undulatory character of light. A light source is placed behind a barrier with two slits (Figure 9.1). As the light waves pass through the slits they will spread, a phenomenon known as diffraction. The result will be two sets of waves that will interfere with each other. The interference can be constructive when the peak of one wave adds to the peak of another wave, or it can be destructive when the peak of a wave encounters a depression. The result is seen on a screen.

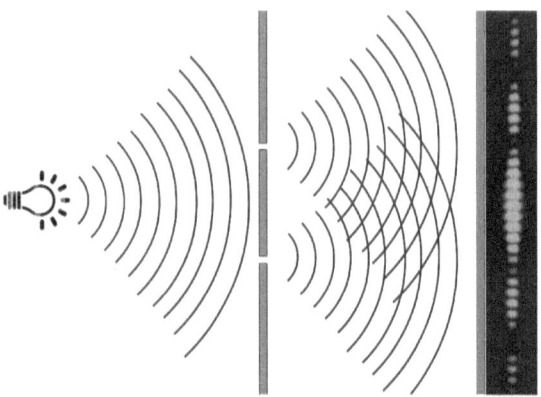

Figure 9.1: Double slit experiment. The interference between the waves causes the striped pattern seen on the right.

Where the interference is constructive, there is a fringe of light, where the interference is destructive, a dark band, forming an interference pattern. Note that if the light was a particle beam such a pattern would not form, we would have only two illuminated areas, behind the slits. And that's how Thomas Young demonstrated the undulatory character of light.

However, as we have seen, light is also made of particles: the quanta of light or photons. And electrons and other particles also behave as waves. We are facing the wave-particle duality, an essential aspect of quantum physics.

The wave-particle duality can be better appreciated through an experiment carried out in 2012 and whose results were published in 2013[1]. The experiment is similar to Young's

---

[1] Roger Bach, Damian Pope, Sy-Hwang Liou e Herman Batelaan. Controlled double-slit electron diffraction. New Journal of Physics, Volume 15, March 2013.

experiment, but instead of using a light source, an electron source was used, and it was conducted in such a way that one electron was emitted at a time. That is, a second electron was emitted only after the first electron reached the screen. Even so, the same pattern of interference appeared as the electrons reached the screen. It is as if the electron, even passing through a single slit, noticed the presence of the other slit and adjusted its trajectory accordingly. It is not possible to predict the path, nor the position that the electron will reach the screen. We have here the two fundamental points that differentiate quantum physics from classical physics. Quantum physics is intrinsically probabilistic and *non-local*[2].

An experiment like this cannot be explained classically, it is necessary to use quantum mechanics for this purpose. But the explanation of what occurs will depend, in this case, on the interpretation of quantum mechanics. We will see the main interpretations in the next chapters.

---

[2]The term nonlocal refers to the possibility that an event in position A will instantly influence position B, even though A and B are far apart.

Chapter 10

The Copenhagen Interpretation

Before we talk about the Copenhagen Interpretation, perhaps we should ask ourselves: Why the need for an interpretation of the theory? Why do discussions about interpretations not arise in classical theories? As an example, let's consider Newton's classical mechanics. Let us suppose that our goal is to determine the future position of a planet. Once we have the position of the planet and other nearby celestial bodies, we can write a set of equations that will predict how the position of the planet varies in time. The accuracy of the result will depend on the accuracy of our knowledge of the initial conditions, as well as possible simplifications. The result may be good or bad, but there is no doubt about what we are predicting: the position of the planet and how this position varies over time. In contrast, quantum mechanics provides us with a wave function and how this function varies over time. The wave function assigns a number to each position in space, with these numbers you can determine probabilities. But what is this wave function? Does it have a physical reality like an electromagnetic wave or is it a theoretical construction? Does it reflect reality or our knowledge of reality? Another aspect to consider is that, in quantum mechanics, if the position of a particle is accurately determined, according to the Uncertainty Principle, this affects its speed, which in turn hinders the prediction of its future position. The very concept of trajectory can be called into question.

Such questions arose as quantum mechanics was being developed. Bohr and Heisenberg were the main ones responsible for what became known as the Copenhagen Interpretation – this name first appeared in 1955 from a series of lectures given by Heisenberg. Accurately presenting the postulates of the Copenhagen Interpretation is an almost impossible task because Bohr's interpretation is not the same as Heisenberg's, nor that of others who contributed to its elaboration. Two texts served as the basis for the presentation of the Copenhagen Interpretation which I give below: the paper *The Quantum Postulate and the Recent Development of Atomic Theory*, published by Bohr in Nature magazine in 1928, and the book *Physics and Philosophy: The Revolution in Modern Science*, published by Heisenberg in 1958.

Our understanding of the world uses classical concepts, such as waves and particles, however, these concepts cannot be applied directly to quantum entities[1], which are neither

---

[1] I use the term quantum entity here to refer to something whose behavior is described by quantum mechanics, avoiding the use of the term particle.

one nor the other. Moreover, the application of these concepts is limited by the Uncertainty Principle. Classic concepts – such as waves and particles, or position and velocity – are complementary in the understanding of quantum phenomena.

Knowledge of quantum phenomena comes from experiments, which, in turn, involve the interaction between quantum entities and an experimental apparatus, classically described. *No matter how much the phenomena transcend the scope of the classical physical explanation, the description of all data must be expressed in classical terms.* states Bohr. What the experiment measures is what it is designed to measure, that is, an experiment can characterize either the wavelike nature or the corpuscular nature of a quantum entity.

There is no point in talking about the behavior of a quantum entity except in the process of measurement. Until the measurement is made the system is a superposition of states, that is, it will be in a combination of possibilities. An electron before reaching the screen – in the double-slit experiment – will be in several positions simultaneously. In Heisenberg's words: *The transition from the possible to the real only occurs in the act of observation.* This transition from the possible to the real is called the collapse of the wave function. Still in the double-slit experiment, before the electron reaches the screen what we have is a superposition of states – with a probability associated to each state. When the electron reaches the screen, there is no more sense in talking about probabilities, the collapse of the wave function occurred.

In essence, the Copenhagen Interpretation states that

it is not possible to attribute properties to the quantum world, as these arise only in interaction with an observer – in the measurement process. In Bohr's words: *There is no quantum world. There is only an abstract quantum physical description. It is wrong to think that the task of physics is to find out how nature is. Physics concerns what we can say about nature.*

The Copenhagen Interpretation can be criticized in many ways. The first problem is to reduce the scope of the theory to the prediction of experimental results. The theory is much more than this. We can use quantum mechanics to explain the nuclear reactions that occur inside stars, or to design materials – such as semiconductors and superconductors – that would be impossible to create according to classical physics. Unless the terms experiments and measurements are used so broadly to lose all meaning, we are not in these cases conducting experiments and measurements, and quantum mechanics is still valid. The emphasis on what is observable is exaggerated. Understanding observable behavior is often impossible without postulating a mechanism that explains this behavior, even if the mechanism cannot be accessed directly. It is interesting to note that the emphasis given to experiments and observables is a late result of the influence of Ernest Mach.

Mach was an Austrian physicist and philosopher of the late 19th and early 20th centuries who published works in the field of optics and developed techniques for measuring the propagation of sound – in his honor, the ratio between the speed of an object and the speed of sound is called the Mach number. Dedicating himself to the philosophy of science, he defended that scientific theories would only be forms of systematizing experimental results and that, therefore, they should abstain from concepts that could not be verified experimentally – for Mach the theories should be only descriptive. Following these ideas, he rejected the concepts of atoms and molecules for not being directly verifiable. He was also a strong opponent of statistical mechanics, a theory which gives a microscopic basis to the results of thermodynamics

The epistemology developed by Mach – and, consequently, the Copenhagen Interpretation – if followed to the letter, would be harmful to scientific development by barring the search for causal explanations that were based on non-observables. Richard Feynmann – one of the greatest physicists of the 20th century – writes in his Lectures on Physics[2]: *It is always good to know which ideas cannot be checked directly, but it is not necessary to remove them all. It is not true that we can pursue science completely by using only those concepts which are directly subject to experiment.*

The lack of care in using the terms measurement and observer also deserves criticism. Although this is not the position advocated by Bohr, the Copenhagen Interpretation led to the perception that any experiment would only have a defined result in the presence of an observer. This brought a strong element of subjectivity, which is completely unrelated to physics and totally unnecessary to quantum mechanics. Bohr's position was that measurement was any interaction between quantum and classical entities. Such a position is found in many textbooks on quantum mechanics, as in the classic Landau and Lifshitz Theoretical Physics Course:

---

[2]Feynman, Leighton e Sands, *The Feynman Lectures on Physics* Vol. III. New Millennium ed., 2011.

Measurement is any interaction between classical and quantum objects[3]. But when should an object be classically described? Where is the borderline between classic and quantum? If every experiment were described from the quantum point of view, would there be no measurement? These questions are not only left unanswered, but they also are not even asked, even though this is one of the fundamental elements of the interpretation. Also, according to the Copenhagen Interpretation, it is in the measurement process that the collapse of the wave function occurs, *the transition from the possible to the real*. What we have before the measurement is a superposition of states. This leads to absurdities, as is clear from the fictional experience proposed by Schrödinger as a criticism, and which became known as Schrödinger's Cat. The experience is simple, imagine a room where there is a cat. In the room, besides the cat, there is a certain amount of radioactive material and a Geiger counter[4], which detects whether or not the material has disintegrated. A device connected to the counter releases a lethal dose of poison if the radioactive material disintegrates. Suppose that by doing the calculations, using quantum mechanics, we come to the conclusion that the probability of disintegration occurring is fifty percent – the system would be in a superposition of states. According to the Copenhagen Interpretation, the cat would be simultaneously alive and dead, until an act of observation – a measurement – caused the collapse of the wave function. As entertaining as the idea

---

[3]L. Landau e E. Lifshitz. *Mecânica Quântica - Teoria não relativista*. Editora Mir, Moscou, 1985.

[4]The Geiger counter (also called Geiger-Müller counter) is a device designed to measure some types of radiation.

of a zombie cat may be, this is not something that suits a scientific theory very well.

Despite the criticism, the Copenhagen Interpretation ended up becoming the standard interpretation of quantum mechanics. It is fair, therefore, to ask why. This is due to several factors. John von Neumann[5] published the book *Mathematical Foundations of Quantum Mechanics* in 1932, in which he demonstrated that any interpretation of quantum mechanics based on hidden variables was impossible, that is, any interpretation in which the probabilities arose from an underlying deterministic theory. The demonstration, in fact, had a flaw and did not prevent interpretations using non-local hidden variables, but this remained unnoticed. Any attempt at an interpretation using hidden variables was soon dismissed, as von Neumann had already demonstrated that this was impossible. Another factor that influenced the adoption of the Copenhagen Interpretation was Bohr's personality. Considered at the time a giant of Einstein's stature, Bohr was the mentor of the generation of physicists who created quantum mechanics. Their admiration for Bohr can be measured by the words of John Archibald Wheeler[6]: *nothing has done more to convince me that there once existed friends of mankind with the human wisdom of Confucius and Buddha, Jesus and Pericles, Erasmus and Lincoln, than walks and talks under the beech trees of Klampenborg Forest with Niels Bohr.* With such a level

---

[5] John von Neumann was a mathematician, physicist, computer scientist, and engineer. He is considered one of the greatest scientists of the 20th century.

[6] John Archibald Wheeler, American theoretical physicist who worked with Bohr in Copenhagen in the thirties.

of admiration, it is no surprise that Bohr's interpretation has been accepted. From this early core in Copenhagen, it spread through textbooks and became orthodoxy.

The founding physicists who were not disciples of Bohr remained, for the most part, critical of the Copenhagen Interpretation, among them: Einstein and Schrödinger.

Chapter 11

De Broglie-Bohm's Theory

While Bohr and Heisenberg developed the Copenhagen Interpretation, de Broigle devised an alternative, which at the time was called Pilot Wave Theory and, later, De Broglie-Bohn Theory. When presenting his theory, at the 1927 Solvay Conference[1], de Broglie was met with strong criticism from the advocates of the Copenhagen Interpretation, in particular, Wolfgang Pauli. Unable to respond to all criticisms, de Broglie ended up abandoning his theory for more than two decades, only resuming it after it was rediscovered by David Bohm.

David Bohm was a researcher working at Princeton University in 1951 when he published the book *Quantum The-*

---

[1]The Solvay Conferences are a series of scientific conferences that brought together the most renowned scientists of the 20th century. They were held at the Solvay International Institute of Physics and Chemistry, located in Brussels.

ory[2], in which he presented quantum mechanics in a didactic way, including the Copenhagen Interpretation, which he accepted. Shortly after the publication, he was invited to a conversation with Einstein – who was a professor at the Institute for Advanced Studies of the university. Einstein presented to Bohm his criticisms of the Copenhagen Interpretation, emphasizing the subjective character of the interpretation. The criticism had an impact and after a few weeks, Bohm developed a new interpretation for quantum mechanics. What Bohm had actually done was to reinvent De Broglie's Pilot Wave Theory, which he didn't know.

According to the De Broglie-Bohn Theory, what we previously called quantum entities are particles, whose movement is dictated by an equation derived from Schrödinger's equation. The particles follow well-defined paths, guided by waves, and the probabilistic character of the theory arises from the ignorance of the initial conditions of the motion. If we consider again the double-slit experiment, what we have now are particles that follow trajectories like those presented in Figure 11.1. The trajectory that the particle will follow is something that will depend on the initial conditions, but once we have the final position, we can reconstruct the path that the particle traveled. The paths the particle can follow are obtained from the wave equation, therefore, phenomena like interference appear naturally. The theory is deterministic, with hidden variables, exactly the type of theory that supposedly von Neumann would have demonstrated to be impossible.

In the fifties, the United States lived under the political persecution of Macarthism. David Bohm, who had been

---

[2] Quantum Theory, David Bohm. Prentice-Hall, 1951.

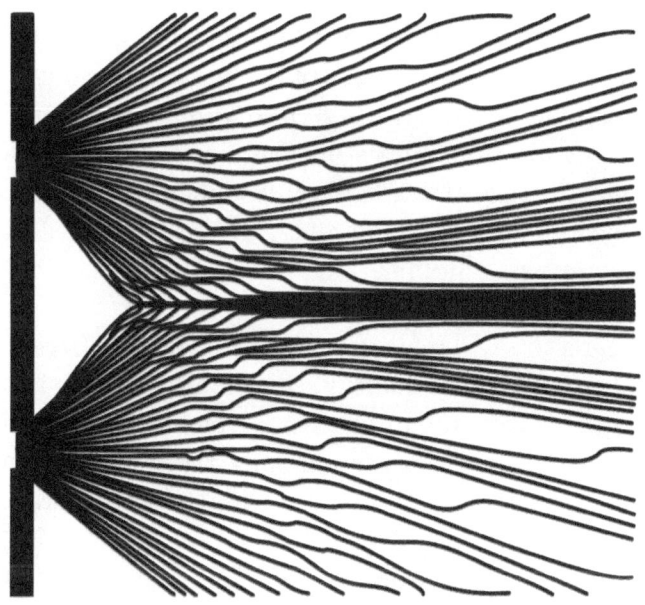

Figure 11.1: Possible paths of a particle in the double slit experiment according to the De Broglie-Bohm Theory. (By File:Doppelspalt.jpg: Opasson / *derivative work Malyszkz - File:Doppelspalt.jpg, Public Domain, https://commons.wikimedia.org/w/index.php?curid=16971210)

affiliated with the Communist Party, had to leave the country. With Einstein's support, he went to Brazil, where he stayed for four years at the University of São Paulo. Away from the great centers of scientific development, it became even more difficult to defend his theory, which was received quite negatively. We can get an idea of how the theory was received by observing what happened at Princeton. Oppenheimer[3] gathered a group of physicists to discuss Bohm's theory and, after a few hours of discussion without finding any error, concluded: *If we cannot refute Bohm, we must agree to ignore him.* And so it was done. For a while, the theory caught the attention of philosophers of science rather than physicists. Lately, however, interest in it has grown and the number of publications on the theory has increased significantly.

Needless to say, De Broglie-Bohm's Theory offers a much more intuitive view of quantum mechanics. Its critics, however, argue that it adds unnecessary elements without producing new results. This is partially true, there is the addition of trajectories. But at the same time, references to measurements and observers become unnecessary. The fact that it is a non-local theory does not help its acceptance either, although it is not possible to escape the non-locality of quantum mechanics, whatever the interpretation[4]. And lastly, a relevant observation about De Broglie-Bohm's Theory is that it is based on Schrödinger's equation, and there is not yet a relativistic version of the theory. In other words, as attractive as the theory is, in its current configuration it is incomplete.

---

[3]Julius Robert Oppenheimer, the theoretical physicist who led the Manhattan Project, was the director of the Institute for Advanced Studies at Princeton University.

[4]Experiments of the type proposed by Einstein, Podolsky and Rosen, demonstrate the nonlocality of the theory.

Chapter 12

The Many Worlds Interpretation

Hugh Everett III was a Ph.D. student at Princeton in 1957 when he presented his thesis on Many Worlds Interpretation. Schrödinger had presented similar ideas a few years earlier, but they were forgotten, just as Everett's thesis probably would have been forgotten had it not been for the 1973 publication of *The Many-Worlds Interpretation of Quantum Mechanics* by Bryce S. DeWitt and Neill Graham. Everett's interpretation has gradually gained popularity and currently has many followers, especially among cosmologists and physicists who deal with quantum computing. There is more than one version of the Many Worlds Interpretation, the one we will present is the original one, developed by Everett in his doctoral thesis.

According to Everett, when describing the temporal evolution of a quantum system, the Copenhagen Interpretation allows for two types of change:

I - Continuous and deterministic changes described by Schrödinger's equation.

II - Discontinuous changes, when there is an interaction between a quantum system and a classical system in the measurement process – the collapse of the wave function.

Suppose that the objective now is to describe the evolution of a quantum system interacting with a classical system, that is, in a measurement process. According to the Copenhagen Interpretation, this would be impossible, because a discontinuous change cannot be described by Schrödinger's equation. This is especially problematic when one wishes to apply quantum mechanics to describe the evolution of the universe, in cosmology.

Everett's solution was to discard the collapse of the wave function and say that the various alternative solutions in Schrödinger's equation take place in different worlds. Returning to the double-slit experiment, we would say that all the various positions where the particle can reach the screen occur simultaneously in diverse worlds, but in each world, the particle follows a unique path. Thus, whenever we have several possibilities in a quantum system – say, the radioactive decay of a nucleus of a uranium atom occurring in a distant galaxy – the world divides and all the alternatives occur. The worlds divide and divide into a potentially infinite myriad.

There is no inconsistency between the Many Worlds Interpretation and the formalism of quantum mechanics. There is also nothing in the formalism that leads to this interpretation. And since the various worlds are independent, no experiment can contradict the interpretation. On the other hand, and this is a personal impression, whenever I hear something about the Many Worlds Interpretation, I get the impression that we are trying to adapt reality to the theory and not the opposite.

Chapter 13

Final Remarks

I would like to finish with some of my impressions and experience studying quantum mechanics. At the end of the nineties, there was a series of lectures with the philosopher Mário Bunge at the Federal University of Santa Catarina. It was in these lectures that I had my first contact with quantum mechanics and the difficulties of its interpretation. Bunge – who, before turning to philosophy, was a theoretical physicist – was a fierce critic of the Copenhagen Interpretation. I was vaccinated against the paradoxes of orthodox interpretation before even knowing the mathematical formalism of theory. When I started to study quantum mechanics more seriously, starting my master's degree, what surprised me most was the use – and misuse – of language in textbooks. Where one could say, for example, *the probability of a particle being in such a position*, one would say *the probability of a particle being measured in such a position*. The reference to measures seemed to be essential, even when out of context – when no measure, not even hypothetical, would be taken. Professors, on the other hand, had almost no interest in interpretations. Like most physicists educated in Copenhagen orthodoxy, they adopted the posture of ignoring interpretations and making calculations. At the time, possibly motivated by the lectures, I read several books by Mario Bunge and also Karl Popper. From these readings I got the impression that it is possible to elaborate realistic and consistent interpretations of quantum mechanics, just by accepting that reality can be intrinsically non-local and possibly non-deterministic.

It was also at that time that what we might call quantum mysticism – the use of the obscurity of interpretations of quantum mechanics to justify extravagances – emerged. Nothing against the mystics – whoever wants to believe, let him believe – but there is nothing in theory to justify them. There is even less to justify pseudo-sciences and charlatanism that goes from alternative quantum therapies to self-help books. I'm sorry if my vision is a bit crude, that's the way the hard sciences are.

Regarding De Broglie-Bohm's Theory, I had never read much about it before I started writing this book. I liked what I read and I believe it may be the best alternative if we can develop a relativistic version of the theory. While this version doesn't come, I think we should accept the theory for what it is: a probabilistic theory. Particles exist, regardless of being observed, and have trajectories, even if they cannot be determined with infinite precision – the Uncertainty Principle does not prohibit trajectories, only their precise determination. The dynamics of fundamental particles follow non-local probabilistic rules, given by the Schrödinger's or Dirac's equation. For now, that's what we have.

Bibliography

BOOKS

Baggott, Jim. *The Quantum Story: A history in 40 moments.* – Oxford University Press, 2011.

Becker, Adam. *What Is Real?: The Unfinished Quest for the Meaning of Quantum Physics.* – Basic Books, 2018.

Bohr, Niels. *Atomic Physics and Human Knowledge.* – Dover Books on Physics, 2010.

Born, Max; Auger, Pierre; Schrödinger, Erwin; Heisenberg, Werner. *Problemas da Física Moderna.* Trad. Gita K. Guinsburg. – Editora Perspectiva, 2000.

Bricmont, Jean. *Making Sense of Quantum Mechanics* – Springer, 2016.

Bunge, Mario (editor). *Quantum Theory and Reality.* – Springer-Verlag, 1967.

DeWitt, Bryce S. e Graham, Neill (editores). *The Many-Worlds In terpretation of Quantum Mechanics.* Princeton University Press, 1973.

Einstein, Albert. *Out of My Later Years.* – Philosophical Library/Open Road, 2015.

Farmelo, Graham. *The Strangest Man: The Hidden Life of Paul Dirac, Mystic of the Atom.* – Basic Books, 2009.

Feynman, Richard P., Leighton, Robert B. and Sands, Matthew. *The Feynman Lectures on Physics*, vol 3. Basic Books, 2011.

Freire Jr., Olival; Pessoa Jr., Osvaldo, Bromberg, Joan Lisa (Organizadores). *Teoria quântica: estudos históricos e implicações culturais.* – EDUEPB/Livraria da Física, 2011.

Gamow, George. *Thirty Years That Shook Physics: The Story of Quantum Theory.* – Dover Publications, 1985.

Gribbin, John. *The Quantum Mystery.* – eBook Kindle, 2015.

Gribbin, John. *Six Impossible Things – The Mystery of the Quantum World.* – MIT Press, 2019.

Heisenberg, Werner. *Physics and Philosophy: The Revolution in Modern Science.* – Penguin Books, 1990

Isaacson, Walter. *Einstein: His Life and Universe.* – Simon & Schuster, 2008.

Kragh, Helge. *Dirac: A scientific Biography.* – Cambridge University Press, 1990.

Kragh, Helge. *Simply Dirac (Great Lives)* – Simply Charly, 2016.

Moore, Walter J. *A Life of Erwin Schrödinger.* – Cambridge University Press, 1994.

Nussenzveig, Moysés H. *Curso de Física Básica 4: Ótica, Relatividade, Física Quântica.* – Editora Edgard Blücher, 1998. (in portuguese)

Pais, Abraham. *The Genius of Science: A portrait gallery of twentieth-century physicists.* – Oxford University Press, 2000.

Pais, Abraham. *Subtle is the Lord: The Science and the Life of Albert Einstein.* – Oxford Universit Press, 2005.

Planck, Max. *Autobiografia científica e outros ensaios.*

Org. César Benjamin; translation by Estela dos Santos Abreu. – Contraponto, 2012. (in portuguese)

Piza, Antônio F. R. de Toledo. *Schrödinger & Heisenberg: A Física além do comum.* – Odysseus Editora, 2003. (in portuguese)

Popper, Karl. *A Teoria dos Quanta e o Cisma na Física.* Translation by Nuno Ferreira da Fonseca. – Publicações Dom Quixote, 1992. (in portuguese)

Popper, Karl. *The Logic of Scientific Discovery.* – Routledge, 2005.

Rosa, Pedro Sérgio. *Louis de Broglie e as ondas de matéria.* Dissertação (mestrado) – Universidade Estadual de Campinas, Instituto de Física Gleb Wataghin. Campinas, SP, 2004. (in portuguese)

Serway, Raymond A., Moses, Clement J. and Moyer, Curt A. *Modern Physics* – Thomson Learning, 2005.

Smolin, Lee. *Einstein's Unfinished Revolution: The Search for What Lies Beyond the Quantum.* – Penguin Press, 2019.

PAPERS

Bohr, Niels. The Quantum Postulate and the Recent Development of Atomic Theory. *Nature*, 1928.

Bunge, Mario. Survey of the Interpretations of Quantum Mechanics. *American Journal of Physics*, 1956.

Bunge, Mario. Quantons are Quaint but Basic and Real, and the Quantum Theory Explains Much but not Everything: Reply to my Commentators. *Sciense & Education*, 2003.

Campbell, John. Ernest Rutherford and his path to the nuclear atom. *Australian Physics*, vol. 48, 2011.

Commins, Eugene D. Electron Spin and Its History. *Annual Review of Nuclear and Particle Science*, 2012

Hughes, J. Rutherford, Radioactivity and the Origins of Nuclear Physics. *Journal of Physics: Conference Series*, 2012.

Kallio-Tamminen, Tarja. The Copenhagen Interpretation of Quantum Mechanics and the Question of Causality. Infinity, Causality and Determinism – *Cosmological Enterprices and Their Preconditions* – Colloquim in Helsinki, 2000.

MacGregor, I. J. D. Ernest Rutherford his genius shaped our modern world. *Europhysics news*, 2011.

MacKinnon, Edward. Heisenberg, Models, and the Rise of Matrix Mechanics. *Historical Studies in the Physical Sciences*, Vol. 8, 1977.

Martens, H. The uncertainty principle. Eindhoven: Technische Universiteit Eindhoven, 1991.

Popper, Karl R. Quantum Mechanics without "The Observer". *Studies in the Foundations Methodology and Philosophy of Science*, vol. 2. Springer Verlag, 1967.

Robitaille, Pierre-Marie. Max Karl Ernst Ludwig Planck: (1858 –1947) *Progress In Physics*, 2007.

Singh, Rajinder. Max Planck and the genesis of the energy quanta in historical context. *Current science*, 2008.

Schleich, Wolfgang P., Greenberger, Daniel M., Kobe Donald and Scully, Marlan O. Schrödinger equation revisited. *Proceedings of the National Academy of Sciences of the United States of America*, 2013.

Trimmer, John. The Present Situation in Quantum Mechanics: A Translation of Schrödinger's "Cat Paradox" Paper. *Proceedings of the American Philosophical Society*, 1980.

www.ingramcontent.com/pod-product-compliance
Lightning Source LLC
Chambersburg PA
CBHW030445220526
45464CB00006B/2419